# · A · DOLPHIN · GOES · TO · SCHOOL ·

# · A · DOLPHIN · GOES · TO · SCHOOL ·

## The Story of Squirt, a Trained Dolphin

•

ELIZABETH SIMPSON SMITH

ILLUSTRATED BY
TED LEWIN

WILLIAM MORROW AND COMPANY, INC. NEW YORK

Printed in the United States of America.

1 2 3 4 5 6 7 8 9 10

Library of Congress Cataloging-in-Publication Data
Smith, Elizabeth Simpson.
A dolphin goes to school.
Includes index.
Summary: Describes how dolphins are trained to perform tricks through the experiences of Squirt, a bottlenose dolphin whose successful training led to appearances in marine shows all across the country.
1. Dolphins—Training—Juvenile literature.
2. Squirt (Dolphin)—Juvenile literature. [1. Dolphins—Training. 2. Squirt (Dolphin)] I. Lewin, Ted, ill.
II. Title.
GV1831.D65S65 1986 636.08'88 85-28407
ISBN 0-688-04815-3
ISBN 0-688-04816-1 (lib. bdg.)

*This book is dedicated to Roy Dellinger,*
*Squirt's trainer and friend*

# ACKNOWLEDGMENTS

*The author sincerely thanks Roy Dellinger, Squirt's trainer, for permitting her to tell the story of Squirt and his training and for checking the manuscript for accuracy; Ralph Quinlan of Quinlan Marine Attractions, the school in which Squirt was trained, for allowing her to visit and observe dolphin training; Ellie F. Roche, Permit Specialist, and Charles A. Ovaretz, Branch Chief, both of the Southeast Regional Office, National Marine Fisheries Service, St. Petersburg, Florida, for information regarding federal regulations; and Diane Edwards Caccia, former dolphin trainer, for supplying valuable photographs for research.*

# ·CONTENTS·

Dolphins and Dolphin Schools 1

1 Good-bye to the Ocean 8

2 First Day at School 22

3 Learning to Retrieve 28

4 What a Show-Off! 35

5 Squirt Defies Nature 40

6 A Baby Dolphin Is Born 46

7 Squirt Grows Smarter and Smarter 54

8 Squirt Flunks His First Exam 61

9 Graduation 70

Index 83

# • DOLPHINS • AND • DOLPHIN • SCHOOLS •

If you have ever seen dolphins perform in marine shows, you have probably seen them jump through hoops or shake their trainers' hands. They may have turned somersaults or caught a football while swimming backward. Perhaps they played a game of baseball, hitting base runs and even arguing with the umpire.

Dolphins can perform so many tricks in such an expert manner that they look as though they have been doing these things all their lives. It is true that dolphins are extremely smart. Their brains are larger even than the human brain, sometimes as much as one and a half times as big. But dolphins were not born knowing how to perform tricks. They had to learn to do these things just as you and I had to learn to add

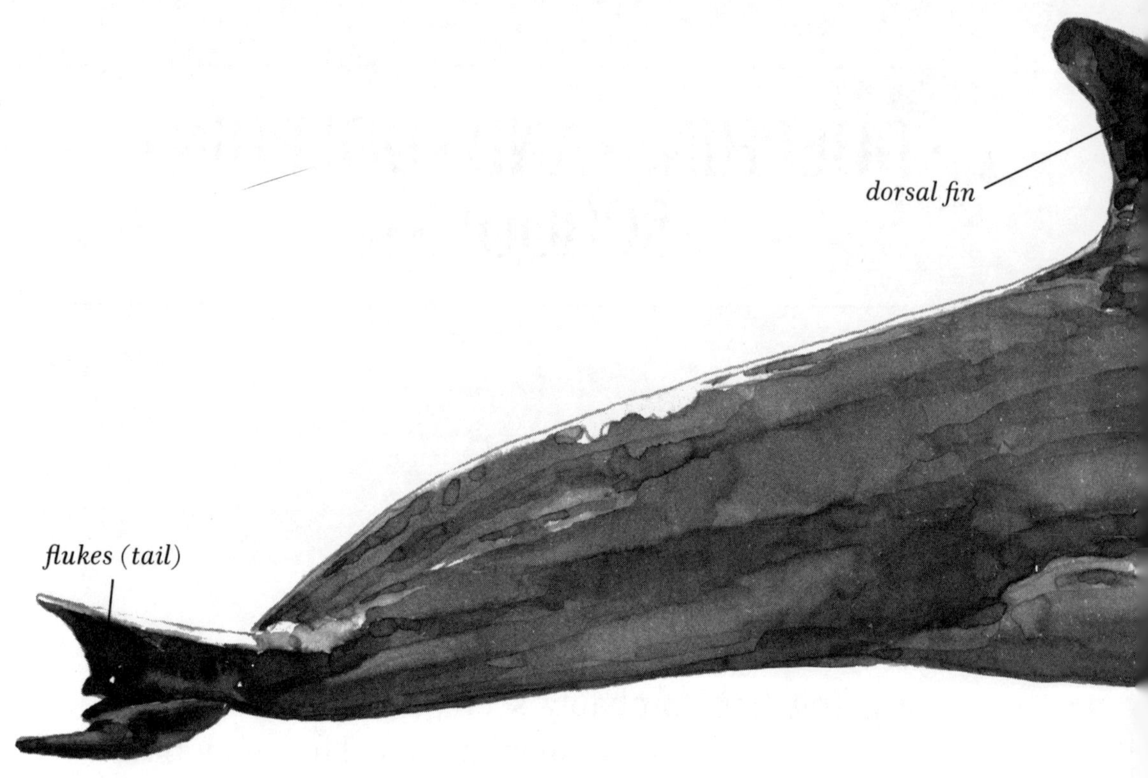


numbers together or to spell our names. They went to school.

All dolphin schools are located on land. Although dolphins born in captivity can be trained at schools, most dolphin students come directly from the wild. This means the dolphins had to leave their homes in the ocean and learn to live in saltwater tanks at the schools. It was an entirely new and strange life for them. No longer were they able to splash around in the sea, to trail fishing boats, or to swim as far away from shore as they wanted. Also, they had to leave behind all the dolphins they were used to swimming

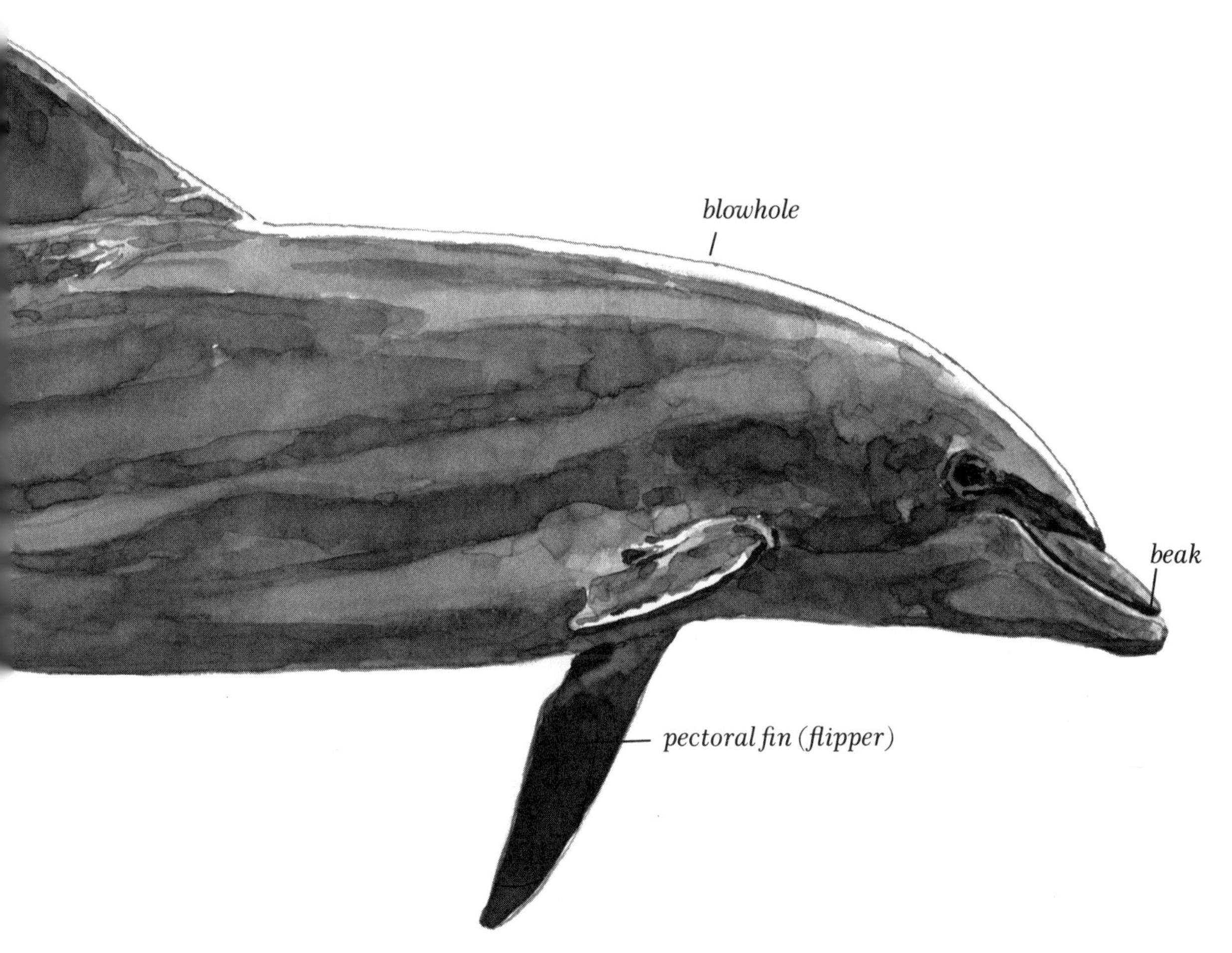


with. Now they had to obey the commands of people and live in small tanks instead of having the whole ocean for a playground. Everything was suddenly very different. It was no doubt a very frightening time.

Dolphin training schools are owned by private citizens. Before a school can be opened, it first must be inspected by an agent from the United States government to be sure the tanks are large enough, the quality of water is right, and the filter system works properly. Once the school is open, it must be reinspected at least once a year. The trainers at a dolphin school also must be approved by the government.

In addition, there are federal regulations governing the people who catch dolphins for schools. These people must get permits by applying in writing to the government, giving their backgrounds, experience with marine mammals, and their reasons for wanting to collect dolphins. When they are approved, they become known as designated dolphin collectors. There are about twelve collectors in the United States. An average of fifty dolphins is brought into captivity each year for either training or research. All of them are collected off the coast of the southeastern United States, which is the only authorized area for collecting dolphins.

The government would not want dolphins to be mistreated or harmed just to bring pleasure to people. To prevent this, there are regulations set forth in the Animal Welfare Act and the Marine Mammal Protection Act. These rules are enforced by the United States Department of Agriculture and by the National Marine Fisheries Service.

One of these federal regulations states that any dolphin brought from the sea must be at least eighteen months old. At this age the dolphin would no longer be nursing milk from its mother and would be able to find its own food. Age is judged by the length of the dolphin. If the dolphin measures at least six feet long, it is considered eighteen months old or older. Except for this regulation, the age of the dolphin does not matter a great deal. Older, or larger, dolphins train as easily as young ones do. And as long as a dolphin remains healthy, it can perform until it dies. Most researchers estimate that dolphins live for twenty to twenty-five years.

Another federal regulation states that if a dolphin does not adjust to captivity—for instance, if it refuses to eat or appears unhappy—within thirty days, the collector may return it to the same area of the sea and collect another one. This rarely happens, however, for most dolphins quickly adjust to their new life among people. A dolphin that has adjusted to being fed and cared for cannot be returned to the ocean, for it would not know how to live in the wild again.

Dolphins trained for shows usually belong to the species called bottle-nosed because they have been found to be the easiest to train. When you look at them, you can see why they are called bottle-nosed, for their long, thin noses do resemble bottles. You also can see that the curves of their mouths make them appear to be smiling all the time. This is pleasing to the audience.

Dolphins must learn a number of tricks before they are ready to appear in a marine show. The tricks are called behaviors or disciplines. Some dolphins learn twenty or more behaviors. This usually takes six months. Of course, not all dolphins learn the same behaviors, nor do they learn them at the same speed. Although there are differences in the way dolphins are trained, all trainers use a reward system of some sort. They have found that dolphins will not respond to punishment, but they will respond to praise and rewards.

After a dolphin knows at least eighteen behaviors, it is ready to perform in a show. Sometimes a trained dolphin is sold to a theme park or a marine show. The trainer goes along with the dolphin to its new home and works with the new trainer until the dolphin learns to trust and obey its new teacher. This usually takes three weeks.

At other times the dolphin simply is contracted to appear at a certain show for a period of time, just as a singer or dancer may perform at many different auditoriums. Between shows the dolphin returns to its school.

All performing dolphins must keep in practice. Even

after they have become expert at a routine, they must continue their training almost every day, just as human performers do. They also must update their acts and learn new behaviors. Although they seem to enjoy doing their tricks, they must work hard.

This is the true story of one bottle-nosed dolphin that was captured and trained and eventually appeared in marine shows all across the country.

# 1 · GOOD-BYE · TO · THE · OCEAN ·

Squirt, a bottle-nosed dolphin, lived in the Gulf of Mexico off the coast of Port St. Joe in northern Florida. It was an excellent place for dolphins to live. There was always plenty of seafood to satisfy their big appetites and to keep them strong and healthy. Almost every day commercial fishing boats went out to sea from the large fishing piers at Port St. Joe. Dolphins followed the boats and gobbled up the fish that were churned up in the wakes. Then, as if to pay for their food, they circled the boats and herded large schools of fish right into the nets. This made the fishermen happy and their jobs much easier.

Since Florida is located in the southern part of the United States, the water along its coastline is usually warm, even in winter. It is a popular spot for dolphin collectors, for they know that bottle-nosed dolphins prefer warmer water than some of the other species do.

The January day that Squirt was pulled from the sea, however, was cold and gray. An early-morning storm had whipped up the waves, and the sun was hidden behind the clouds. The three men who arrived from the dolphin school—the owner, his assistant, and a trainer—were bundled up in winter clothes. Because the trainer would have to go into the water, he wore a wet suit, the kind that deep-sea divers wear. The suit was made of rubber so it would keep the water out and would remain warm inside from body heat. He carried a stick exactly six feet long for measuring the dolphins and a barrel of live saltwater fish for bait. The men rented a fishing boat equipped with a small motorboat, a heavy fishing net, a canvas sling large enough to hold a dolphin, and a creel to lower or hoist the sling. The owner of the boat served as pilot.

At first the pilot guided the boat close to shore, where dolphins usually stay. The other men stood near the rails and used binoculars to search the water. Dolphins are different from most sea creatures. Members of the whale family, they are mammals and must come to the surface to breathe through the blowholes on top of their heads. A flick of their flukes, or tails, brings them to the surface. They arch gracefully above water as they puff out used air and draw in fresh. The blowholes then snap shut to keep water from entering, and the dolphins again dip under the waves. The rhythmic curving above water every few minutes makes them fairly easy to find. But no dolphins appeared.

The pilot nosed the boat about ten miles farther from land, out beyond the breakers, where the surf was smoother. Dolphins are likely to move to calmer waters when the surf near the shore is rough. But the men searched for more than an hour without seeing any sign of them. Finally the trainer spied one dolphin in the distance, then another. The pilot steered the boat closer, and the trainer reached into the barrel and began tossing fish overboard. Suddenly the area was teeming with dolphins, gracefully curving above the water to breathe, then diving underneath to grab one of the fish. The pilot idled the motor and tossed out an anchor.

The assistant climbed into the motorboat and was lowered into the water. The other men tossed one end

of the net to him. He started the motor and slowly made a wide sweep around the larger boat, pulling the net behind him. This way he was able to fence off a circular area measuring ten acres, or about the space of ten football fields. The net was twenty feet wide, with weights attached to its lower edge. The weights pulled the lower edge of the net deep into the water, while the upper edge remained near the surface. This formed a corral similar to the kind used for capturing horses, except this one was in water.

The trainer pulled a life jacket over his wet suit,

grabbed his measuring stick, and jumped overboard. He quickly checked every area of the net to be sure no dolphins were trapped below the surface. If any were, they would drown if they could not get to the surface to breathe.

There were thirty-two dolphins within the corral. They splashed water on the trainer and nudged and bumped against him in a friendly manner. They appeared to be curious but not afraid since they were in their natural habitat. (There is no way to know for sure what a dolphin is thinking, of course, because we can make judgments only in human terms. That is, if the dolphin appears to us to be happy or scared or angry, it is because it is acting the way a human would act under those circumstances. In this case they appeared to be friendly.)

The owner of the school was a designated dolphin collector and had been granted a permit from the United States government to capture six dolphins. He watched from the rail of the boat, picking out the dolphins that had no visible injuries. He also noticed if they seemed spunky and playful, for this probably meant they were in good health. When he saw a dolphin that looked just right, he would shout to the trainer and point out the dolphin to him. The trainer would swim to the dolphin and measure it with his stick. If the dolphin was large enough, the trainer would pat it on its head and talk and play with it until he sensed the dolphin was not afraid. Then he would quickly slip one arm underneath the dolphin and the other arm over its body and grab a pectoral fin, or flipper, in each hand. The dolphin usually tugged and flipped its flukes, but without the use of its pectoral fins it could not break away. When the dolphin grew still, the trainer guided it to the boat. The sling was lowered into the water, and the trainer nosed the dolphin inside. The sling and the dolphin then were hoisted into the boat.

Squirt was the third dolphin captured that day. He at once became the trainer's favorite because he was such a tease. Trainers like to find teasing dolphins because they usually do well in shows. They enjoy having fun and will work hard to get special attention.

Squirt played a sort of hide-and-seek or guess-where-I-am game with the trainer. Just as the man swam toward him, Squirt disappeared. In a moment he resurfaced, this time on the other side of the trainer. When the trainer stuck out his measuring stick, Squirt nipped playfully at it. Then he disappeared and again came up on the other side of the trainer. The next time the dolphin disappeared and resurfaced, the

trainer managed to measure him and to look him over. Fortunately Squirt was a little longer than the stick, so he would be a legal catch. There was a tiny nick in one fluke, probably a shark bite, but not big enough to be noticeable. This time Squirt nudged the man on his left side before disappearing. The trainer looked to his right, expecting the dolphin to play the game again. But Squirt fooled him. He popped up on the same side from which he had disappeared, nudging the trainer as he surfaced. Just as the trainer turned his head, the dolphin clamped his jaws together and squirted a mouthful of water right into the man's face. From that moment the trainer called him Squirt.

As soon as Squirt was lifted aboard the fishing boat, the owner used a violet-colored spray paint to write a number three on Squirt's back. Squirt's number would identify him as the third dolphin caught that day.

Bottle-nosed dolphins range in color from cream to shades of gray. Sometimes there are yellow spots scattered about like huge freckles. Dolphins' skin is very tender and will get air-burned when out of water for too long a time, just as human skin will burn when exposed to sun. To prevent this, the men covered the dolphins with wet flannel blankets. But first each dolphin was examined quickly to see if it was male or female. The government requires that the sex be reported as a matter of record. The female has a longer

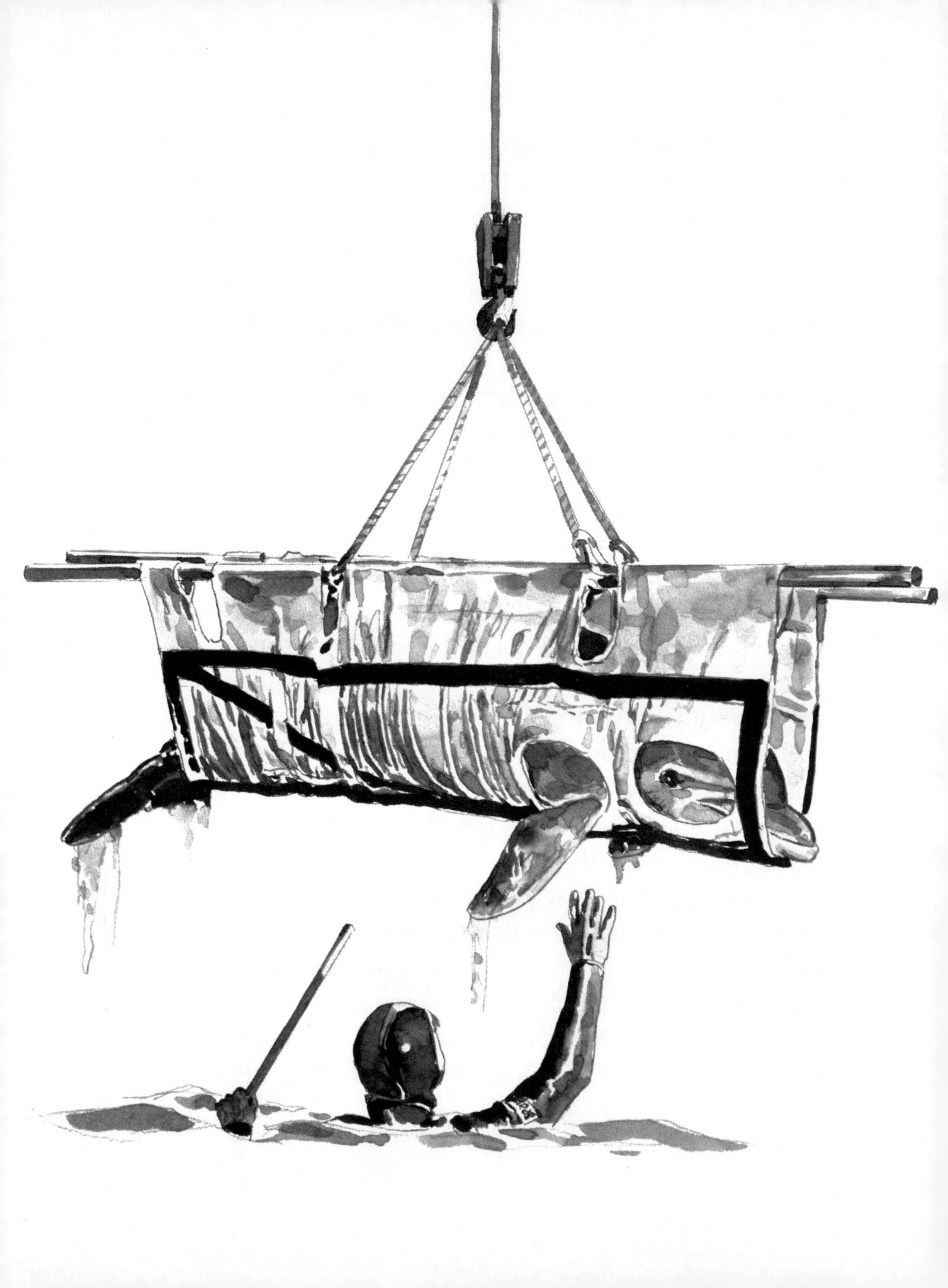

genital slit than the male, long enough to permit the birth of a baby. Both sexes train equally well. Squirt was a male, so he was covered with a blue blanket. Pink blankets were used for females.

While still in his sling, Squirt was placed in a metal holding tank prepared especially for the occasion. The tank was thirty feet wide and six feet deep and half-filled with water. A metal rod ran the length of the sling on two sides. These rods were fitted into special holders at the top of the tank. The canvas sling, with Squirt inside, hung down from these rods into the water. This arrangement made it possible for Squirt to rest comfortably and to breathe while keeping the lower part of his body wet. The upper part, except for the head, was covered with the blanket. Every five minutes a bucket of fresh seawater was splashed over the blanket to keep it wet.

After all six dolphins had been captured, the boat and net were removed from the water, and the trainer and assistant climbed aboard the fishing boat. The journey to shore took more than thirty minutes. During this time the trainer took a sample of blood from each dolphin, using a needle and vial just as a doctor would use on a human. An analysis of the blood would help determine the health of the dolphins.

As the boat bounced across the waves, Squirt and

the other dolphins grew still. They didn't even talk among themselves in little clicking pulses or with their whistle tones, as dolphins often do. Once ashore, they were released from their slings and placed in a waiting pool, an area of ocean that is netted off so they could swim but would not be able to escape. They would remain in the pool for at least twenty-four hours to be sure they were healthy and could make the trip to the school. During this time they were not given any food since eating might make them carsick on their journey.

The blood samples were rushed to a local hospital for humans for laboratory tests. Squirt's tests were perfect, but the number two dolphin was not so lucky. She had an infection and had to be given a shot of medicine in the hip.

During the night the trainer and the assistant took turns going to the pool to check on the dolphins. They found them squealing and whistling to one another noisily.

On the morning of the second day Squirt and his companions again were put into their slings and returned to the holding tanks. The tanks were loaded onto a truck. The trainer gently rubbed zinc oxide into the dolphins' skins to protect them further from air, just as football players use this ointment to protect their faces from sunburn. He left a circle free around their eyes and their blowholes.

At noon the two men and the six dolphins departed for North Carolina, where the school was located. The owner drove home alone in a passenger car. The trip was more than four hundred miles. The trainer and the assistant took turns driving. One man always stayed in the back of the truck so he could check the dolphins' breathing, which should be three to five breaths per minute when they are not in deep water, and could keep the blankets wet. Again Squirt and the others grew still and quiet.

During the twelve-hour drive the dolphins still were not being fed. This does not mean that they were hungry, however, for dolphins eat great quantities at one time. They have three stomachs and can store food to digest later as they need it. Sometimes they eat as much as fifty pounds in one meal and can go up to three weeks without eating again.

When they were only thirty minutes away from the school, the men stopped to telephone a veterinarian and ask that he meet them on their arrival. Another trainer also would meet them to help unload the dolphins. The two trainers would divide the dolphins, so each one trained three. Squirt was already spoken for by the trainer who had pulled him from the ocean.

About midnight the truck turned off the highway and rattled down a country road half-hidden by shrubs and trees and closed off by a gate. It came to a stop at a low concrete building with a loading dock at one end. The building, hundreds of miles from an ocean, would be both Squirt's new home and his schoolhouse.

The men lifted the slings onto the dock so the dolphins could be examined again. The veterinarian inserted a cotton swab into Squirt's blowhole to remove a culture and took a blood sample from an area between his flukes and vertebrae. He also weighed and measured him. Squirt was six feet eight inches long and weighed 225 pounds. This means he was quite young. Many adult bottle-nosed dolphins measure at least eight feet long and weigh more than 300 pounds.

The main room of the building had the hollow sound and the smell of an indoor swimming pool. It held six round tanks of salt water, about ten feet in diameter and eight feet deep. As the dolphins were slipped into their tanks, they were given names since the numbers on their backs would fade away. Squirt, of course, had been named already.

That first night two dolphins shared a tank so each would have company. Squirt's partner was Juno, a female much larger than he. Squirt clung to her side.

Juno nudged the younger dolphin affectionately and clicked and whistled to him in a cooing manner, just as a dolphin mother at sea would do.

The trainer stayed for almost an hour after the veterinarian and the assistant had left. He wanted to be sure the dolphins were swimming and breathing properly. Finally, at two o'clock in the morning, he turned out the lights and left Squirt and his companions alone for the first time in their strange new boarding school.

# 2 • FIRST • DAY • AT • SCHOOL •

When the trainer arrived at seven o'clock the next morning he found the dolphins swimming quietly. Squirt still swam close to Juno. The veterinarian came by to report that Squirt's culture showed he had a parasite in his blowhole. The men used a net to lift Squirt onto a stretcher, strapped him to a table so he would not fall off, and gave him a shot of medicine in his hip. Again they used a wet flannel blanket to protect Squirt's skin from air burn. When Squirt was returned to the tank, he raced to Juno's side.

The trainer tossed a fish into the center of Squirt's tank. He was careful to cut off the head of the fish first since the fish was not alive. Dolphins eat fresh fish from the sea after killing them by biting off their heads. They would be suspicious of a dead fish with its head still intact. Squirt, however, showed no interest in eating and let the fish sink to the bottom of the tank. Since dolphins are trained by using food as bait, Squirt could not begin his schooling until he started eating again.

At this time the trainer set up two charts, one for Squirt's medical record and one for his diet. From now on he would fill in the charts every day and add any special notes of importance. That day he wrote on the diet chart, "Will not eat."

By the third day Squirt was showing a little more independence by swimming apart from Juno. The trainer also noticed that Squirt had begun talking to Juno underwater. He could see little bubbles of "words" emerge from Squirt's mouth and ripple to the top of the water. Dolphins often communicate with one another with a variety of sounds—barks, grunts, creaking-door noises, and whistles. The trainer hoped that Squirt's signals to Juno meant he was feeling more at home. But the young dolphin still refused to eat, even though Juno now was eating several pounds a day. Nor would Squirt swim to the side of the tank where the trainer stood.

The fifth day, however, was different. When the trainer tossed a fish into the tank, Squirt dived for it, gobbled it up, and surfaced for more. That day the

trainer threw three pounds of fish before Squirt had his fill. The next day Squirt ate five pounds. The following day he ate eight pounds. After another week he was up to his full diet, which is about fifteen pounds for a dolphin living in captivity. Now Squirt was ready for his first classroom lesson.

The next morning at feeding time Juno was removed from the tank and Squirt was left alone. The tank, like a school desk, would be the area where Squirt would learn. It was equipped with a platform that extended about four feet over the water and was large enough to hold one or two people. Metal guardrails ran along three sides of the platform. The trainer could stand on the platform and lean into the rails for balance and support. This would leave his arms free

when he eventually began teaching the dolphin from a standing position.

The first lessons, though, were at water level. They began with the trainer tossing one fish into the tank. Squirt dived, ate it, and came up for more. This time the trainer lay on his belly on the platform and held the beheaded fish by its tail between his thumb and forefinger exactly at water level. Squirt circled the tank, keeping the man in view, and waited for the tossed fish. The trainer, however, would not toss it. Nor would he call or coax the dolphin with his voice. All commands would be made by only one thing: the hand that held the fish.

Squirt seemed puzzled and backed off to the opposite side of the tank, where he remained for several minutes. Then he began working his way toward the center of the tank. Each time he dived and surfaced he moved a little closer to the platform. The trainer wiggled the fish slightly to tempt Squirt. Eventually the dolphin worked his way timidly to the trainer's side. Then he suddenly took a deep dive, surfaced, and grabbed the fish from the trainer's hand, all in one big motion. Squirt had learned his first behavior!

The trainer at once whistled between his teeth in one short, shrill blast. This was the first time he had whistled in Squirt's presence. From now on this whistle, along with a second fish, would be Squirt's rewards for performing properly. Eventually the whistle would bring even more pleasure to Squirt than eating the rewarding fish. It would be like getting a good report card or an *A+* on a test.

In grabbing the fish, Squirt had nipped the trainer's finger with his teeth and had drawn blood. But the man continued without pausing. This time he held the fish a little farther up the tail, still at water level. Squirt hesitated only briefly, then lunged for the fish and grabbed it from the trainer's hand.

Each time the trainer held another fish he moved his hand farther and farther away from the tail. This made Squirt come closer and closer to the man's hand. The trainer had one goal in mind for this first series of lessons: to touch Squirt on his nose when the dolphin grabbed a fish. He wanted Squirt to become familiar with the human touch. This would make it easier for Squirt to trust the trainer and do as he was asked when they began to do tricks that demanded more skill and body contact.

Since dolphins quickly become bored, the first class lasted for only ten minutes. By the end of the lesson Squirt was still lunging for the fish and jerking it from the trainer's hand. During the next lesson the trainer would try to get him to take the fish more slowly. This would help Squirt learn to move gracefully and would be important when he performed in shows.

The trainer fed Squirt properly and ended the class until the next morning.

# 3 •LEARNING•TO•RETRIEVE•

Now that Squirt was in training he lived alone in his tank. He knew, though, that other dolphins were nearby. Sometimes he stood up on his flukes and peeked into their tanks.

The dolphins often talked among themselves in whistles and clicks, just as those living in the ocean do. At times they made longer noises like a cat or a dog—a screech, a yelp, or a bark. If they were underwater, the sound was much softer, but the trainer could tell the length of the sound by the number of bubbles that drifted to the surface.

The trainer, of course, did not know how much talking and peeking and playing went on in his absence. But often in the mornings he could hear squeals and clicks before he unlocked the door. When he opened the door, the sounds would stop, but fat, telltale bubbles floated on the surface of each tank. This pleased Squirt's trainer, for it indicated that the dolphins were happy and enjoying themselves.

The morning after Squirt's first lesson the trainer found the young dolphin hovering near the center of the tank, watching the man cautiously. When the trainer set down his bucket of fish, Squirt cut through the water and perched himself just below the training platform.

"Hello, Squirt," the trainer called. He squatted beside the tank and patted Squirt's head. "Good Squirt. Good boy."

He picked up a headless fish, placed his fingers at the tip end near the water, and held out his hand. Squirt flipped above the water, took the fish with a big splash, and allowed the trainer to touch his nose. Good. Squirt had remembered what he had been taught the day before. He was ready for lesson two!

The trainer whistled immediately and tossed a second fish into the pool. After a few minutes Squirt began to move more smoothly and to pause slightly after taking the fish, as if he were enjoying the touch of the trainer's hand. His leaps grew more graceful, and he splashed less water over the edge of the tank.

However, he still nipped the trainer's hand with his

sharp teeth. Dolphins have between eighty and a hundred pointed teeth, which easily can puncture human skin. The trainer's fingers began to look as full of pricks as a pincushion, and little circles of blood popped out on their tips.

As that morning's lessons progressed, the trainer began to toss the rewarding fish off to the right. This coaxed Squirt to swim in the same pattern each time: to approach from the left, swim to the right, and circle back. This lesson was important. When Squirt appeared in shows, he would be working with other trainers. The new trainer then could be told that Squirt performed in a left-to-right manner so there would be no confusion.

Squirt worked in ten-minute segments. Then the trainer called a recess. While Squirt played or rested, the trainer moved to the next tank to train another dolphin.

At first Squirt didn't seem to mind when the trainer moved away, but by midmorning he appeared to be jealous. He rose on his flukes, found the trainer with his eyes, and squealed for attention. This pleased the trainer, for it meant that Squirt had not lost interest. It also meant, he hoped, that Squirt was growing to care for him. Each time the trainer returned, Squirt seemed happy to see him. He lingered close by the

platform and let the trainer pat his head and rub his body. This was a good sign.

After a few hours Squirt was gliding smoothly through the water, rising gracefully to snap the fish, and swimming off to the right with hardly a splash. He seldom nipped the trainer's fingers anymore. By midday, however, he began splashing noisily and squirting water in the trainer's face. The man took this as a clue that Squirt was bored with repeating the same behaviors, so he moved on to the next lesson.

This time Squirt would be working with something he had never even seen before—a soft rubber ball as big as a softball and as yellow as a lemon. Now the training technique would be slightly different.

Squatting on the platform, the trainer held a piece of fish in his right hand as usual. In his left hand he held the ball. As Squirt sprang from the water to claim his fish, the trainer gently touched the dolphin's nose with the ball. As usual he whistled and tossed another fish off to the right.

The dolphin seemed puzzled, but he circled back for more. This time the trainer held the ball in his right hand without a fish. As Squirt jumped, the trainer touched the dolphin's nose with the ball. Using his left hand, he picked up a fish from the bucket and tossed it to the right.

When the trainer felt that Squirt was familiar with the ball, he began another technique. This time he pressed the ball into Squirt's mouth to force it open a little, whistled, and tossed a fish. After a few more tries Squirt began opening his mouth wider and wider. Finally the trainer was able to place the ball inside Squirt's mouth before the dolphin swam away. Squirt, of course, dropped the ball to catch his fish reward and left it floating on the water. The trainer leaned over, scooped up the ball, and had it ready for Squirt's next approach. In this way he began to show Squirt that the ball was not to be left in the water.

The lessons proceeded all day long. After each recess the trainer opened the class with a review of the behaviors Squirt had already learned, starting with a piece of fish and progressing to the ball. Each class would begin this way for the rest of Squirt's life. A review helps a dolphin remember the lessons it has learned, just as it helps a boy or a girl.

In the afternoon the trainer again changed his lesson plans. Now when Squirt approached him, the trainer dropped the ball into the water but refused to whistle or toss a fish. Squirt circled the ball, nudged it with his nose, and paused. But there still was no whistle and no fish reward. Finally Squirt seemed to realize that he would get no reward until he picked the ball from the water and returned it to the trainer's hand. This is called retrieving.

It would be many days before Squirt would become good at retrieving. When he was, the trainer then would begin standing on the platform, holding his hand higher and higher, and throwing the ball farther and farther across the pool. Squirt would learn to pick up the ball at the opposite end of the tank, dive deeply, swim across underwater, then leap higher than the trainer's head to return the ball. Eventually the trainer would stand on a ladder and Squirt would leap fifteen or twenty feet while an audience clapped and cheered. But that would be many months away. Squirt was still in kindergarten.

After classes were over the second day, the trainer wrote on Squirt's chart: "Ate fifteen pounds. Learned to retrieve ball from pool side. Lets me pat and rub him. Seems well and happy."

# •WHAT•A•SHOW-OFF!• 4

Dolphins like to swim in water that is between sixty and eighty degrees Fahrenheit, just as humans do. The temperature in Squirt's tank was a pleasant seventy to seventy-five degrees. Water was pumped in through a filter system, just as it is in swimming pools, so there was a constant flow.

The school had to turn fresh water into salt water so it would be heavy enough to support Squirt's weight. In fresh water a dolphin would sink too deep to be able to breathe regularly. Salt was blown into the water through long hoses attached to a power unit on a truck parked just outside the door. Squirt's tank held about 42,000 gallons of water. At first 10,500 pounds of salt were blown in. After that salt was added as needed.

Every day the trainer checked to make sure that the

pool held the right amount of chlorine. For this he used the same kind of test kit that is used to check swimming pools for people.

One day he found a fungus growing on Squirt's skin. When he ran tests, he found that too much chlorine had killed not only the bad bacteria but also the good bacteria that Squirt needed to stay healthy. The trainer grew cultures of good bacteria in an incubator and added them to the water. Then the amount of chlorine was adjusted. But Squirt had to be given several shots of antibiotics before his skin was healed.

After this incident Squirt was given a big dose of vitamins every day to keep him well. He never realized he was taking them, though, because the vitamins were slipped inside the fish he ate.

Every three months a long refrigerated trailer truck came bouncing down the road and backed up to the door of the school. On those days Squirt and the other dolphins had a holiday. School was closed while both trainers unloaded forty thousand pounds of saltwater fish—blue runners, Boston mackerel, and Spanish mackerel. The fish were frozen and packed in fifty-pound paper cartons. It took all day to unload the cartons and stack them in the freezer.

The trainer sliced open a carton every morning, took out sixteen pounds of fish, cut them up, inserted a vitamin in one slice, and put them all into a bucket. This is the fish he rewarded Squirt with during class time. At the end of the day the trainer fed Squirt all he wanted from what was left in the bucket. In this way the trainer could write on the chart exactly how many pounds of fish Squirt ate every day.

The trainer also wrote on the chart about Squirt's bowel movements, or stools, as they are called. He could tell from their appearance if the dolphin had an upset stomach and if he needed medicine. He also could tell when Squirt didn't feel good by the way he lay around lazily instead of playing. A dolphin doesn't sleep the way a human sleeps because it must always rise to the surface to breathe through the blowhole on top of its head. Yet it can't float above the water because its skin would get air-burned. When a dolphin rests, it hovers just below the waterline. Every four to six minutes a flick of its flukes brings it to the surface to catch a breath of air.

Squirt usually felt good and was playful even when he wasn't taking a lesson. Occasionally a visitor was permitted at the school. Squirt always noticed someone new and showed off by doing his tricks. If the visitor laughed or applauded, the dolphin repeated his act, then rushed to the side of the pool so the person could pat his head.

One day Squirt became so eager for a pat on the head that he pushed himself out of the water and landed on the side of the pool, right at the feet of his startled visitor. He had to wiggle his flukes in order to work his way back into the water before he became air-burned. Later Squirt would learn a behavior called dry-docking in which he would come out of the water on purpose. But on that day he was too frightened from his accident. It was no time to learn a new behavior.

# 5 •SQUIRT•DEFIES•NATURE•

Squirt's next lesson was probably the scariest one he ever would have to learn. It was the hoop trick, and because it defies a dolphin's natural instinct for self-preservation, it would be extremely frightening to Squirt.

Dolphins have a different system of hearing from people. Although dolphins have ears, they hear through their lower jaws. At night or in dark or muddy water dolphins even can use their jaws to see. To do this, the dolphin sends out a series of evenly spaced clicks. The sound races through the water until it bumps into something solid—a coral reef or a shark the dolphin would want to avoid or a fish it would like to eat. Then the echo bounces back as a signal to the dolphin. From the returning signal a dolphin can create a "picture" of the object and decide how to deal with it.

This sending and receiving of sound is called echolocation or sonar. It is such a good system that the United States Navy uses the same principle in submarines to locate underwater objects. It also is used in radar for airplanes.

Many scientists believe that dolphins once were land animals and used their ears to hear. Similarities between the bodies of dolphins and land animals point to the truth of this belief. For instance, the dolphin has an ear on each side of its head just as land animals do. It also has hips like four-legged animals and a set of three stomachs like a cow. Unlike fish, dolphins are warm-blooded. They don't have scales like fish, they don't swim like fish, nor do they lay eggs.

These scientists believe that dolphins probably lived on beaches so they could eat the lush green plants growing near the water. Then they began going out into the water to catch fish for food. Eventually, over a long period of time, they moved into the sea to live.

Underwater, however, their ears were not good enough to hear the muted sounds of the ocean. So dolphins developed a hearing system in their jaws, and their ears grew smaller and smaller. Today a dolphin's ears are quite tiny for such a large mammal.

In the tank Squirt's sonar system was useless, for his clicking sounds simply bounced off the sides of the small pool. He didn't need it anyway, for he could see quite well in the clear water. But his instinct had taught him to fear an enclosure. He knew if he swam inside such an enclosure, he could become trapped in it underwater. That would mean death to a dolphin just as it would to a human.

When the trainer first held a bright blue hoop in his hand, Squirt's sonar detected an enclosure, and his instinct told him the hoop would encircle his body. He raced to the other side of the pool, as far from the hoop as he could get. He hovered near the edge like a stalled motorboat, staring at the trainer.

The trainer understood Squirt's fear and was very gentle. He cooed to him in familiar words and coaxed him back with a bite of fish. He patted Squirt's head and rubbed his shoulders. When Squirt seemed calm again, the trainer lay on his stomach and held the hoop underwater with his left hand. In his right hand he held a piece of fish between the hoop and the edge of the tank. To get the fish, Squirt would have to pass all the way through the circle.

The sight of the hoop was still too frightening to the dolphin. Squirt waited awhile, then cut through the water and swerved to the right without even touching the hoop. At the opposite end of the tank he again hovered and stared, while the trainer called and coaxed.

On the next try Squirt came close enough to touch the hoop with his nose but refused to pass through. As he swam away, he looked back over his shoulder, awaiting the whistle and the fish. But the trainer made no move.

Again Squirt made a swipe by the hoop, close enough for a touch. But there was no whistle to reward him, no fish to eat.

Now Squirt realized he would have to swim all the way through the hoop before he would be rewarded. Yet if he obeyed his teacher, he would have to turn his back on a lesson he had learned long ago from nature, a lesson that probably had saved his life many times in the sea. What a dilemma!

The trainer faced a dilemma, too. If Squirt decided to go through the hoop, it meant he was placing great trust in his teacher and would be able to learn many other tricks. But if Squirt failed, he probably would not be able to graduate from school. Instead of performing in shows, he would be put into a petting pool for the rest of his life. When dolphins fail to learn enough tricks or if they grow too sick to perform, they are placed in a petting pool at a park or recreation

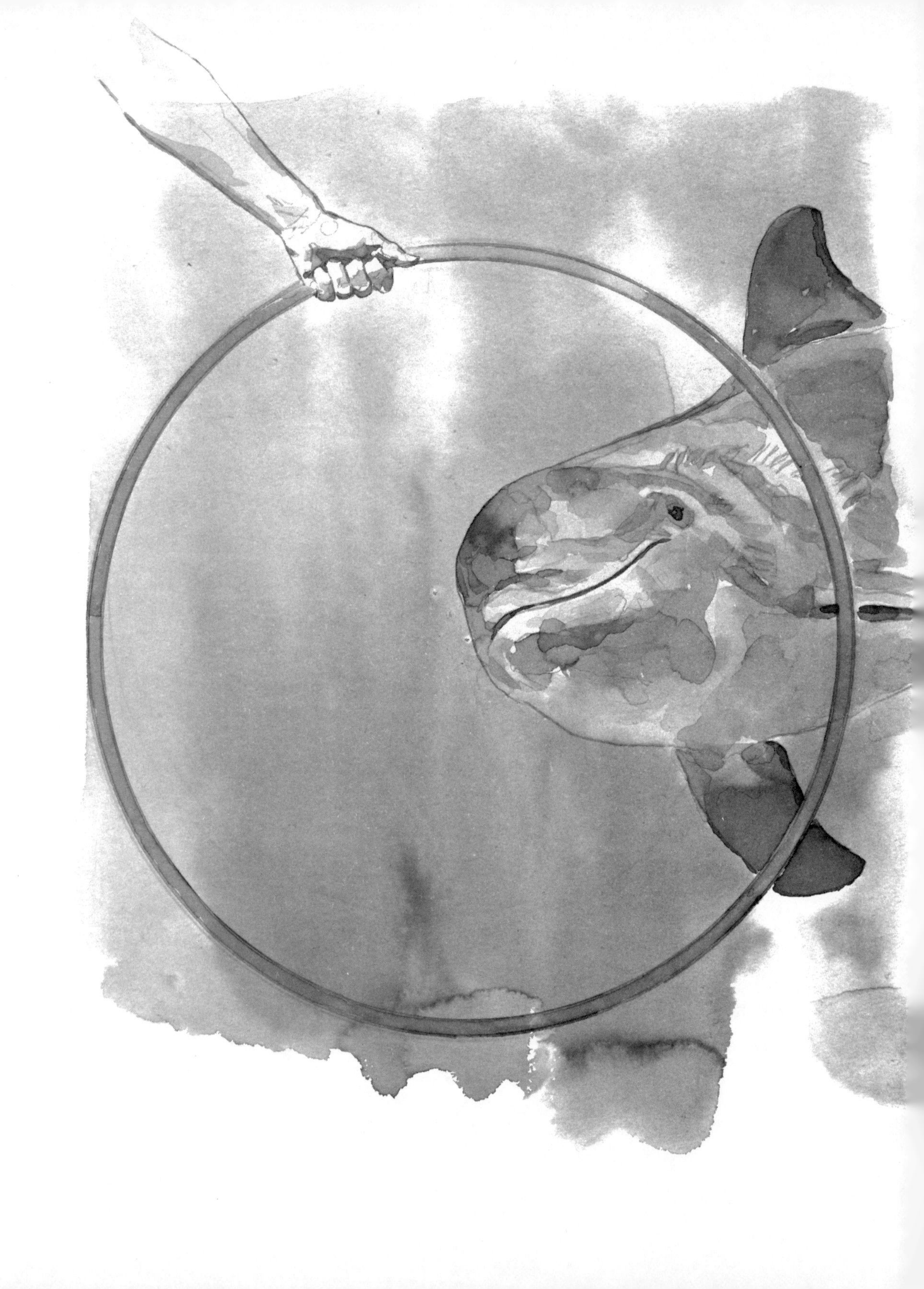

center. There they are fed and cared for properly, and visitors can observe and pat them. It isn't a bad life, for the dolphins enjoy the attention. But life in a petting pool would be too lazy for a spunky dolphin such as Squirt.

The trainer had grown fond of Squirt and certainly did not want him to fail. Perhaps he should call a recess, the man thought, and save the hoop trick for a later lesson.

Suddenly Squirt sprang through the water at the opposite side of the tank, took a breath, dived, and disappeared. In one smooth motion he glided underwater the full length of the pool and swam straight through the center of the hoop with hardly a splash.

Squirt would certainly get an *A+* that day!

# 6 •A•BABY•DOLPHIN•IS•BORN•

One morning the trainer found Squirt rearing up on his flukes like an angry racehorse. He screeched excitedly and stared toward the next tank, where Lena lived.

The trainer set down the bucket of fish, but Squirt paid no attention. He sprang up on his flukes again, turned toward the other dolphin, and squawked like a blue jay.

By then the other trainer had reached Lena's tank. "Come help," he yelled, dropping his bucket with a clatter.

Lena was having trouble staying afloat, a sure sign that she was sick and needed help. She kept sinking below the surface, and it took great effort for her to rise again. If Lena were still in the sea, she would be sending out distress signals, a series of clicks and whistles. Then any nearby dolphin would rush to her aid like a rescue squad. The rescuer would swim under Lena and push her to the surface every time she needed to breathe. Other dolphins also would hear the signal and swarm in to help. They would form a circle to fight off sharks and to take turns holding Lena up to breathe. This is the way dolphins seem to respond to cries for help.

Even in the tank Lena had no doubt been sending out signals, and Squirt had probably been receiving them. But there was no way he could go to her aid.

The trainers were puzzled. Lena had seemed quite healthy and in good spirits the evening before. The other dolphins were well, and there were no viruses making the rounds. Then their eyes were drawn to something startling. A small fluke was sticking out from Lena's genital slit.

"Glory be!" her trainer exclaimed. "Lena is having a baby."

This was alarming news and made the trainers sad. It takes about twelve months for a baby dolphin to develop. Lena had been in captivity for only a few months, so she was already pregnant when she was captured that day in the Gulf near Port St. Joe. If the men who caught her had known this, they would have left her in the sea so she could have raised her baby naturally.

Now Lena would have to drop out of training for eighteen months while she nursed her baby. More important, giving birth to a baby in captivity is quite different from giving birth at sea.

Normally an expectant mother would have picked out another female dolphin to serve as her special friend for the last six months of her pregnancy. The friend would swim beside her and help her stay afloat when she became heavy with the weight of the growing baby. The friend would help the mother-to-be find extra food so she could nourish her fetus. And when it came time for the baby to be born, the friend would be there to push the mother up for air when she needed it.

The friend would help the newborn baby, too. A dolphin is born fluke first because the birth process can take up to two hours. If a baby were born head first, it would drown. After the baby drops free from its mother, it is often unable to breathe on its own. The friend could push the baby to the top to keep it from drowning.

Without a friend at hand Lena seemed frightened. Both trainers jumped into the water and swam to Lena's side. They patted her head and gently rubbed

her back. They joined hands underneath her body and lifted her above water when she needed to breathe.

Meanwhile, the baby's flukes became completely visible and the lower part of its torso was beginning to appear. When this happened, Lena grew calm and made little yips of sound like a contented cat. In less than thirty minutes the full baby appeared.

A baby dolphin should be able to see, hear, and swim at birth, but Lena's baby dropped deep into the water, pulling away from its mother. Lena dived, swam beneath the baby, and pushed it up for air.

In only a few minutes Lena would need to nurse her baby. Her milk was stored in little pouches, or mammae, on each side of the genital slit. A dolphin baby cannot nurse by sucking with its lips as a human does. If it did, it would suck in too much water and would drown. Dolphins use a system that works like a water pistol. An infant swims beside its mother and nudges her near one of the mammae. The mother then knows her baby is in position to nurse. She tightens the muscles around the mammae and squirts the milk directly into her baby's mouth.

Lena stayed close by, but her baby showed no interest in nursing. The trainers helped by gently pushing the baby close to the mammae, but even this did not work.

The baby felt small and light, no more than fifteen or twenty pounds. It was barely two feet long. A full-term dolphin should weigh twenty-five pounds or more and measure at least three feet. The trainers knew then that the baby was premature. It was born too early, before it was fully developed, and it did not have strength to swim or eat. No matter how many times the trainers pushed it to the surface, the dolphin always dropped back down again. Finally the baby stopped breathing altogether.

The trainers put the dead baby onto a stretcher and removed it from the tank. The veterinarian would be called in to examine the body carefully. Perhaps he would learn some new information about dolphins that would help prevent this from happening in the future.

Lena watched carefully all the while but did not try to interfere. She seemed to sense the trainers were doing the right thing. The men knew that Lena was no doubt saddened by her loss, so they watched her closely, fed her well, and showed her lots of attention. In a few days they would keep her busy again learning new behaviors like the other dolphins at the school.

Meanwhile, Squirt had been flipping up on his flukes and watching the activity like a nosy cat. Most of the time he made friendly clicking and whistling noises, which must have been comforting to Lena.

Toward the end, after the trainers had failed to get the baby to breathe and swim, Squirt grew quiet.

When the trainer returned to Squirt's tank, he found the dolphin hovering at the opposite side, as far away from the training platform as he could get. The man suspected that Squirt was confused and perhaps even saddened. Since Squirt and Lena had been exchanging sounds, it is possible that Lena had told him about her baby and its death.

The trainer decided to act as if it were a normal school day. He opened class as usual by beginning with the first behavior Squirt had learned. The dolphin slowly swam over and took a bite of fish from the man's hand. The trainer moved to the platform and held up another piece. Squirt made a lazy leap, barely high enough to grab the fish. The trainer took one step backward and refused to whistle. This indicated that he was displeased with Squirt's performance. Ordinarily the dolphin would try much harder after such a scolding, but today his heart did not seem to be in it. His leap was again disappointing.

The trainer set down his bucket of fish and walked away. When a dolphin is not in the mood for school and refuses to do its best, it is better not to force it to perform for its food. Otherwise, it will expect to be rewarded even if it does sloppy work.

The trainer spent the rest of the morning teaching the other dolphins that were in tanks across the room from Lena and did not seem upset. When he returned in the afternoon, he found no change in Squirt. The dolphin tried a few low leaps, but they were nothing like the six-foot heights he already was capable of jumping.

The trainer stooped by the side of the pool, rubbed Squirt's skin affectionately, and talked to him. Squirt appeared hungry, so the trainer tossed him fish after fish, as long as Squirt would eat. Then he declared a holiday and went off to the laboratory to write on the charts the events of that day.

# 7 • SQUIRT • GROWS • • SMARTER • AND • SMARTER •

In fewer than three months of school Squirt was performing incredible tricks. He could toss a basketball through a hoop. He could catch a football and kick it for the extra point. He could shake hands and turn somersaults. He could tailwalk on his flukes all the way across the tank. He could wave bye-bye. He could even sing.

Now he was living in a world of humans and doing humanlike things. Yet only a few months earlier he had never even seen a human, nor had he known about tanks or balls or hoops.

At this point Squirt couldn't do all the behaviors perfectly, of course. He would have to learn to leap higher and higher, to sing louder, to work faster and more gracefully. And he still had to learn to perform with other dolphins. He would learn at least two new behaviors a year. In time he would invent special tricks on his own. Like other trained dolphins, Squirt would have to study the rest of his performing life. For him this would mean returning to the school in North Carolina whenever he wasn't appearing in shows. Often he would spend all winter in school because many of the parks that give marine shows are open only in summer.

Squirt had learned each behavior the same way, with the whistle-and-reward system. Often learning one behavior had made the next one easier. Jumping over a pole, for instance, was similar to jumping through a hoop. By now Squirt was springing fourteen feet high for both the pole and the hoop. Soon it would be sixteen feet, then eighteen. Eventually he would jump twenty-three feet, higher even than some two-story buildings.

Squirt learned to tailwalk by following a piece of fish in the trainer's hand as the man backed away.

To teach Squirt the somersault, the trainer attached a small ball to the end of a pole. He swung the pole in a wide circle, half above water and half below. Squirt learned to touch his nose to the ball and to follow it around the circle, into the water and back out again. Next he learned to circle in a beautiful one-and

a-half somersault and reenter the water with the grace of an Olympic diver.

Squirt learned to add a special feature to this trick. As he reentered the pool, he gave his flukes a fast flip that sent water sailing out of the tank like a small hurricane. This would drench the first few rows of spectators and was sure to bring roars of laughter.

Squirt had learned tricks with a regular size basketball and football just as he had with the little yellow softball. He could catch a basketball in his mouth, swim to the other side of the tank, and, using his flukes, flip the ball out into the audience. When he played football, he stood on his flukes to catch the ball in his mouth, dived, swam underwater across the pool, then flipped it with his tail for a completed pass.

Learning to kick the extra point was trickier. It involved dry-docking, or coming all the way out of the water onto the side of the pool. Dry-docking is one of the hardest tricks for a dolphin to learn and requires much courage. A dolphin knows instinctively that it will get air-burned if it stays out of the water for more than ten or fifteen minutes. Yet the dolphin must go against its own nature, just as Squirt did with the hoop, if it is to please its trainer.

To teach Squirt dry-docking, the trainer used a soft foam rubber mat with a slick vinyl coating. He held the mat underwater and rewarded Squirt each time he touched it with his nose. Gradually the trainer lifted the mat higher and higher until it was completely out of the water. Eventually he placed the mat flat on the floor beside the tank and coaxed Squirt to climb up out of the water and lie on it. When it was time to get back into the tank, all Squirt had to do was wiggle his hips and slide tail first over the slick vinyl coating.

To kick the extra point after a touchdown, Squirt tailwalked the ball across the pool, dry-docked, and then flipped the football over a goalpost that stood three feet above the mat.

For basketball Squirt pushed the ball across the pool, then bounced it off his nose and into a hoop mounted on top of a pole. To teach this trick, the trainer placed the basketball on top of Squirt's head and pushed both the ball and the dolphin's head into the water. This made Squirt mad, and he began pushing back. Each time Squirt pushed the ball back into the trainer's hand he received a whistle and a reward. The trainer then began tossing the ball into the basket without whistling or rewarding Squirt. Squirt didn't have to read a rule book to understand that he would have to toss his own baskets or he'd get no reward.

Dolphin noises are made through their blowholes instead of their mouths. Although most of these noises

are not very appealing or melodic, Squirt did learn to sing a song of sorts. The trainer taught him to sing by touching his nose with a fish, then jerking the fish away before Squirt could grab it. This made the dolphin mad enough to squawk. Each time he squawked the trainer whistled and rewarded him. When Squirt squawked louder, he got a bigger piece of fish. When he squawked high notes, his rewards were even greater. Soon the squawks became a song. When it came time to sing, Squirt pulled his upper body out of the water and stood just under the platform, with his tail in the water. As Squirt sang, the trainer waved his arms back and forth as though he were directing a great choir.

On some days Squirt was in a bad mood, but most of the time he seemed cheerful, whistling or squealing merrily at class time. He worked hard to please his trainer, a good sign. He would probably work hard to please an audience, too. All in all, the school was quite pleased with him.

# 8 • SQUIRT • FLUNKS • • HIS • FIRST • EXAM •

In late spring Squirt got his first big break in show business. The owners of an amusement park in Virginia asked the school to send them its best-trained student. The dolphin was to appear in the first marine show of the season at the end of May and to remain all summer. It was a great opportunity, and Squirt was chosen.

For the next ten days Squirt worked especially hard. He and his trainer ran through all the behaviors just as if they were performing in public. They began with a friendly handshake. Then they worked with the softball, the pole, hoop, the football, and the basketball. The show would end with a bye-bye as Squirt lay on his side and waved farewell with the flipper protruding above water.

The trainer added extra effects to make people laugh and enjoy themselves even more. He taught Squirt to splash water out into the audience at certain times. He taught him to surprise the audience by repeating a behavior when the audience thought he had finished. He taught him to disappear underwater and then to pop up in some unexpected spot.

Squirt went through his paces like a professional performer. He seemed to have more fun than ever before, and he liked clowning around. The trainer grew confident that Squirt would be a star and the show would be a great success.

On the day he was to leave for the amusement park, Squirt was netted, mounted in a sling, and lifted onto a gurney. He was examined thoroughly by the veterinarian, then covered with a wet blanket and rolled outside. Squirt squealed joyfully in the fresh air and sunshine he had not seen for months.

The crane lifted Squirt onto the truck. The trainers attached his sling inside the tank, already half-filled with water. Now the tank was rigged with a battery-operated sprinkler system that automatically sprayed water. It was no longer necessary for one of the trainers to pour water from a bucket to keep Squirt's skin wet during the trip.

Once again Squirt set out on a journey that would

take him even farther from his home in the sea. This time the trainer and the owner of the school were his traveling companions. They made sure that he always was breathing properly and looked comfortable. As they traveled, Squirt grew quiet, keeping a watchful eye pinned to any activity within view.

Five hours later Squirt and his companions arrived at the amusement park. He was lifted from the truck and again examined by a veterinarian. Then he was placed in an outdoor training tank similar to the one at the school and allowed to relax for the rest of the day. He swam slowly around the tank, his eyes alert.

Meanwhile, Squirt's trainer met with another trainer at the park and wrote a script for the dolphin show. Since there would be six shows a day, the trainers would take turns handling Squirt. While one trainer was performing with him, the other would act as master of ceremonies and announce the show through a microphone.

Squirt would have many more things to learn. At the school he and the trainer had worked together without many words. The trainer would always say, "Good morning, Squirt," when he opened school for the day. And he'd say, "How're you feeling, boy?" when he rubbed the dolphin's back. When Squirt did a behavior especially well, the trainer would cry, "That's my pal," or, "Good boy." But usually they had worked in silence. Now he would have to work in noisier surroundings. The show would have to be announced so the audience would know what to expect. When the master of ceremonies boomed into a microphone, "Squirt will now jump through the hoop," the audience would probably think it was a voice command to the dolphin. Actually the words would mean nothing to him. Squirt would continue to work through hand signals from the trainer the way he had learned. But the noise would be an added distraction.

The next day Squirt seemed happy to be back at work. He was so enthusiastic that sometimes he went directly to the next behavior without even eating the rewarding fish. A few days later the park's trainer began working with him, too. This was the first time Squirt ever had worked for anyone other than his trainer at school. At first he acted shy and unsure. Instead of jumping higher than asked, as he did with his regular trainer, he approached the stranger cautiously and performed only well enough to get his reward. But soon he relaxed and began showing off in his usual way.

For three weeks Squirt and the two trainers worked together just as they would in the show. The trainers made up funny lines to go along with Squirt's antics.

Sometimes workers at the park stopped by to watch. This gave Squirt a chance to work in front of an audience. He splashed playfully when they applauded.

The day before the show was to open Squirt was moved to the tank in the middle of the arena for a final rehearsal. This tank was quite different from Squirt's training tank. It was four times larger. Around its edges ran a platform from which the trainers would work. On all sides stood rows and rows of bleacher seats set up like a small football stadium.

But most unusual of all was the tank itself. This tank was not made of concrete. It was made of an acrylic that looked like clear glass. Nor was it sunk in the ground the way the training tank was. It sat at ground level on top of a concrete slab. This made it possible for the audience to see Squirt from all angles, even when he dived to the bottom of the pool.

Such a setup made things quite different for Squirt. Now sunlight streamed through the sides of the tank and cast rippling shadows on the clear blue water. It may have reminded Squirt of his real home in the ocean, for he began to play. He popped up through one sunbeam, then dived smoothly through another. He raced from side to side, splashing happily as he enjoyed the added space. But during rehearsal he stayed in the small area surrounding the trainer's platform, just as he had been trained. Not once did he go beyond the boundaries he had been taught.

Finally it was show time, a bright, sunny Saturday on Memorial Day weekend. The gates of the amusement park were opened, and boys and girls, men and women streamed through. Squirt's first performance would be at two o'clock in the afternoon. All morning the public-address system announced over and over, "Get your tickets for the dolphin show."

Squirt, too, seemed caught up in the excitement. He squealed when the trainers came to move him to the arena a half hour before show time. Again he splashed and raced from one edge of the tank to the other.

The show began right on time. The trainer called Squirt to his side, touched his nose, and fed him his first reward. The master of ceremonies stepped to the microphone and introduced Squirt and the trainer.

"Squirt will now shake hands with his trainer," the voice boomed. Squirt swam forward, extended his pectoral flipper, and got his reward. There was a burst of applause and a few cheers.

"Squirt will now play softball," the master of ceremonies said.

Squirt swam forward but ignored his trainer's signal. He began backing away. He went deeper and circled the tank. Then he came back to the center and turned slowly around and around. Looking through acrylic was like looking through glass. The dolphin could see something frightening he had never seen before—a throng of people. He could see boys and girls as they unwrapped candy bars or jumped up and down and clapped. He could see men and women as they talked and moved around. He could see the flash of color from their sweaters and T-shirts and balloons. He could hear their shouts and their laughter. Nothing in the sea ever had looked or sounded this way. Even at school and during rehearsals at the park he had seen only a few people at a time. On these occasions the people, like his trainer, had been located outside the pool and at a higher level. Now the spectators appeared to be in the same pool. How strange this world of people had become.

"Squirt will now play ball," the master of ceremonies repeated. Slowly Squirt caught his trainer's signal and obeyed. The audience was delighted, and the trainer breathed a sigh of relief. Squirt made it through the rest of the show, but his heart didn't appear to be in it. His jumps were low and lazy, he was slow, and his timing was bad. Sometimes the trainer had to repeat a signal several times. The crowd cheered heartily, for just to see a dolphin perform is wonderful. They

USA

had no way of knowing that Squirt was only half trying.

"He's sick," the owner of the school said after the first show.

"No, I think it's just stage fright," the trainer said. "He can see those people through that tank. It looks as if they're all in the water with him. He's never seen anything like it in his whole life. He's scared to death."

Nonetheless, the veterinarian was called in. Squirt's health was fine, but his spirits were sagging.

The next day Squirt performed no better. Nor the next. After a week he began losing weight and showing signs of nervousness. Even though the crowd was pleased, his trainer was far from happy with his performance.

There was nothing left to do. The show had to go on. A more experienced dolphin was rushed in from another school, and Squirt was packed up and returned to North Carolina.

Squirt had flunked his first real exam.

# 9 •GRADUATION•

All the way home the trainer wondered and worried about Squirt. Was his problem really stage fright? Had it been caused by the see-through tank, or would it happen anywhere? Would Squirt ever get over it? Or would he never perform again?

Since dolphins are natural performers after they have been trained, stage fright is rarely a problem. Instead, most dolphins respond to cheers and applause, and seldom do they seem to mind the size of the audience. Would Squirt be one of these natural performers?

There was only one way to find the answer.

As soon as the truck arrived back at the school, the trainer unloaded Squirt, who was still very quiet, and put him in his own familiar tank. Quickly the man prepared a bucket of fish and returned to the training platform. This time he found Squirt racing around the tank like a child at a favorite playground. When Squirt saw the trainer, he swam to him and took the first bite of fish from his hand.

The trainer found this encouraging, but still he was filled with doubt. It was possible that Squirt was just hungry, that he had no desire to do his behaviors.

Quickly the trainer gathered a small audience—the owner, the other trainer, and several visitors. He brought out the softball and tossed it into the water. Squirt took the ball in his mouth and returned it to the man, while the spectators cheered. The trainer next brought out the hoop, then the pole. Before he got to the basketball and the football, he already had his answer. Squirt was giving the very best performance of his life—and with an audience!

Squirt's excellent performance convinced the trainer that the dolphin should continue his training. Stage fright would not be a problem at most marine shows since see-through tanks are rare. From now on the school would make certain that only concrete tanks were provided.

Squirt began classes again. So far all of his behaviors had been solo. However, in most marine shows the dolphins perform in pairs or in groups of three or more. Eventually Squirt would work with four or five other dolphins at the same time. Before he could do this, though, he had to learn to work with only one other dolphin.

A few days after his return from Virginia he was moved into a tank with Cocoa, a female who had been taught by the same trainer. This would be Squirt's new home.

At first the two dolphins spent their time pushing each other around and marking out their own private territories to live in. However, soon they became friends and shared their tank equally.

The trainer began with the same behaviors that both Squirt and Cocoa already knew. But now they had to learn to do them together at exactly the same moment and in exactly the same way. It would take weeks and weeks of practice.

Now the trainer used both hands to command. With his right hand he signaled and rewarded Squirt. His left hand was reserved for Cocoa.

What a circus the first few lessons turned out to be! Squirt and Cocoa looked more like clowns than trained dolphins. Sometimes they watched the wrong hand and got their signals crossed. They bumped together in the air. They leaped to different heights. They

jumped through the wrong hoops. And sometimes they missed their hoops altogether.

If one dolphin made a mistake, neither was rewarded. The one who had done a good job then would squawk at the trainer and grow angry at the dolphin that had missed the cue.

Even worse, Squirt and Cocoa grew jealous of each other. When Squirt did a behavior better, Cocoa would try to make him miss on the next try. When Cocoa got more attention, Squirt would live up to his name. He would squirt a mouthful of water at the trainer!

Not all the behaviors were done together. Some would still be solos. Squirt did the high hurdles better, so he would do them alone in a show. Cocoa was a surer shot at basketball, so this would be her solo. But this created another problem, for the dolphin doing the solo got the fish. The trainer worked out a solution, though. When Cocoa tossed a basketball, Squirt was rewarded for staying out of the way. Somehow he understood that his reward this time was for obeying the command not to perform. In the same way Cocoa was rewarded for staying away while Squirt did his high leaps.

After three weeks of hard work and many errors, Squirt and Cocoa were working together smoothly. Once this had happened, the trainer began teaching them new tricks to add sparkle to their show. Each dolphin took turns picking up a small stone off the bottom of the tank with its mouth and returning it to the trainer. At show time, the trainer would take a real wristwatch off his arm to toss into the water instead of a stone. To teach this behavior, the trainer jumped into the tank and retrieved the stone himself to show the dolphins how it was done. As usual, he used the whistle-and-reward system.

The trainer also jumped into the tank to teach the dolphins the hula dance. First he tossed grass skirts (actually they were made of plastic) into the water. Then he taught the dolphins to rise up through the skirts, let them slide down to their hips, then wiggle across the tank Hawaiian-style on their flukes.

Squirt and Cocoa also learned to flip a ball into the stands and wait for someone in the audience to pitch it back.

But their most amazing new behavior was one the trainer borrowed from the United States Navy. Some years ago the Navy used a trained dolphin to carry mail to an underwater laboratory, called Sealab, off the coast of California. The mail was placed inside a plastic bottle, similar to a milk bottle, and the top was screwed back on. Two ends of a rope were tied to the neck of the bottle, making a circle about three feet in diameter. The dolphin poked its head through the circle and did something even the United States Post Office couldn't do. It swam the mail to the underwater lab. This is the same method that Squirt and Cocoa would use in the show to carry articles to the bottom of the tank and back again.

After two more months of practice Squirt got another chance at show business. Although Squirt had flunked his first performance, the trainer thought it was time for another test, this time in a concrete tank. Now that each of the six dolphins at the school had finished six months of training, they all were being sent out on assignments to various marine shows. Squirt and Cocoa were hired to perform at a popular amusement park near Cincinnati, Ohio.

This time Squirt would travel in style. Recently the school had bought an airplane large enough to hold eighteen human passengers. The seats had been removed to make room for the sprinkler tanks, and Squirt and Cocoa were put aboard.

The trainer at the amusement park would handle their show. It would take a couple of weeks for Squirt and Cocoa's new trainer to gain their confidence. But by the end of the second week both the dolphins and the new trainer were working well together. They even devised a new trick or two to tease the audience.

Squirt's regular trainer would stay through the opening show. If all went well, he would return to the school and leave Squirt and Cocoa at the amusement park till the end of summer.

An hour before show time Squirt and Cocoa were moved to the tank in the arena. As the stands filled with spectators, the dolphins began doing tricks on their own. They jumped up on their flukes. They waved their pectoral fins. They splashed and dived. The more the crowd laughed and cheered, the more Squirt and Cocoa cavorted.

This pleased the trainers as they stood backstage and watched. A good warm-up usually indicates that the show will be good, too. Each time Squirt and Cocoa did something especially funny or pleasing, the trainers whistled and threw them pieces of fish. Since the dolphins were not receiving commands from the trainers, they realized they were being rewarded for being clever on their own. That made them try harder and harder. In fact, they were being so funny that the trainers threw a basketball into the tank. Squirt flipped it to Cocoa, and Cocoa bounced it back off her nose. Squirt fielded it with his flukes and sent it sailing out into the audience. A boy caught it and threw it back to Squirt. Squirt flipped it right back into the stands.

At that the trainers relaxed. It was time to begin

the show, and they were sure that Squirt and Cocoa would be a hit.

And they were. Never before had either dolphin done so well. The trainer decided to try one of the new tricks. He put the dry-dock mat beside the pool. Then the master of ceremonies asked if anyone in the audience was celebrating a birthday. A boy and a girl responded. He asked them to come to the tank and stand near the mat.

"Here, Squirt. Here, Cocoa," he called as he signaled with his hand to the dolphins. Squirt and Co-

coa came swimming across the tank and pulled themselves onto the mat. The trainer let the children pat the dolphins and feed them from a barrel of fish. Then Squirt and Cocoa dropped back into the water.

"Did you like doing that?" the trainer asked the children. "Would you like to pet them again?"

"Oh, yes," they answered excitedly. Suddenly Squirt and Cocoa popped up again and startled the children. The dolphins had been trained to reappear, but neither the children nor the audience knew it. The people thought it was a great joke.

The high jumps came near the end of the show. Together Squirt and Cocoa leaped twelve feet with perfect timing. Next came a solo leap by Squirt, who could now jump so high that the trainer had to stand on a cherry picker to hold the fish. Squirt dived deep, came straight up and sailed sixteen feet into the air, the highest jump he had ever taken. While the audience roared with cheers, Squirt and Cocoa lay on their sides and each waved bye-bye with a pectoral fin.

"Ladies and gentlemen," the master of ceremonies announced, "this is the end of our show."

But Squirt wasn't through. The cheering and the clapping seemed to make him want more. He sprang up onto his flukes, stood as long as he could, then fell face forward into the water.

"He took a bow!" the people in the stands cried. "He really took a bow."

The people didn't realize, of course, that Squirt did not know what a bow is. He simply was expressing the way he felt in the only manner he knew. But his action was so convincing that it was added as a special feature. Later, whenever Squirt performed, he took a bow at the end of the show.

The trainer from the school was both happy and sad when he told Squirt good-bye. He was proud of his

student and happy that Squirt had overcome his stage fright. But like most trainers, he had developed a special feeling for the dolphin. He would miss Squirt.

In the fall Squirt would return to the school to brush up on his old tricks and to learn new ones, and the trainer would work with him then. But next spring Squirt again would leave to go to marine shows across the country. From now on he would spend at least half his time away making professional appearances. For Squirt was now a performing star. He had graduated from school.

# • INDEX •

Illustrations are underlined.


air-burn, 15, 18–19, 22, 37, 39, 57
Animal Welfare Act, 4

beak, 3
blowhole, 3, 9, 10, 58

corral, 11, 12

dolphin
  adjustment to captivity, 2–3, 5
  anatomy of, 2–3, 9–10, 17, 19, 30, 40, 41, 47, 50, 58
  birth of baby, 47–48, 49, 50
  bottle-nosed, 5–6, 15
  capture of, 9, 10–13
  communication of, 23, 28
  diet of, 8, 19, 22–23, 24, 36, 37
  government protection of, 4–5, 15
  habitat of, 8–9, 10, 35
  hearing of, 40–41
  intelligence of, 1
  lifespan of, 5
  measuring age of, 5
  nursing of baby, 50
  performing, 1, 5
  sleeping habits of, 37

dolphin (*continued*)
transporting of, 17, 18–19, 62–63, 76
dolphin collector, 4, 12
dorsal fin, 2

echolocation, 41

fluke (tail), 2, 9–10, 13

genital slit, 17, 47, 50

mammae, 50
Marine Mammal Protection Act, 4
marine shows, 1, 6, 55, 61, 71, 76, 82
measuring stick, 9, 12, 13

National Marine Fisheries Service, 4

pectoral fin (flipper), 3, 13
petting pool, 43, 45
Port St. Joe, Florida, 8

Sealab, 75
sling, 9, 13, 16, 17, 18, 62
sonar, 41, 42
Squirt, 14
age of, 20
arrival at school, 20
capture of, 13, 15
fear of enclosure, 42, 43
first performance of, 66–67, 68, 69
health of, 18, 20, 22, 36, 37, 69
naming of, 15
performance with other dolphin, 77–80
playfulness of, 13, 15
reaction to humans, 38–39, 38, 64
refusal to eat, 22–23
relationship with other dolphins, 20–21, 22, 23, 24, 46, 47, 51–52, 72–75, 76
relationship with trainer, 13–14, 27, 29, 30, 31, 32, 34, 43, 45, 53, 60, 71, 80, 82
size of, 20
training of, 25–27, 29–30, 32–34, 42–45, 52, 61–62, 63–65, 72, 74–76

training, 6
  behaviors (disciplines), 6
  familiarizing dolphins
    with props, 26–27, 32–
    34, 42–45
  hand signals, 64, 72, 78
  left-to-right manner, 30
  of pairs of dolphins, 71–
    72, 74–75
  review, 6–7, 33, 55
  reward system, 6, 24–25,
    25, 26, 52, 55
training schools, 2–4
  inspection of, 3–4
  regulation of, 4–5
tricks, 1, 6
  baseball, 1
  basketball, 54, 56, 57, 58,
    77
  dry-docking, 39, 57–58,
    78–79, 78–79
  football, 1, 54, 57, 58
  hoop trick, 1, 40, 42–43,
    44, 45, 55
  hula dance, 75
  leaping, 29, 32, 52, 54,
    55, 73, 80, 81
  retrieving, 33–34, 75
  shaking hands, 1, 54
  singing, 54, 59, 60
  somersault, 1, 54, 55, 57
  tailwalking, 54, 55, 56,
    58, 75
  taking a bow, 80
  waving good-bye, 61, 80

United States Department
  of Agriculture, 4
United States Navy, 41, 75